NIOSH Fire Fighter Fatality Investigation and Prevention Program

Leading Recommendations for Preventing Fire Fighter Fatalities, 1998–2005

by

Marilyn Ridenour

Rebecca S. Noe

Steven L. Proudfoot

J. Scott Jackson

Thomas R. Hales

Tommy N. Baldwin

DEPARTMENT OF HEALTH AND HUMAN SERVICES
Centers for Disease Control and Prevention
National Institute for Occupational Safety and Health

This document is in the public domain and may be freely copied or reprinted.

DISCLAIMER

Mention of any company or product does not constitute endorsement by the National Institute for Occupational Safety and Health (NIOSH). In addition, citations to Web sites external to NIOSH do not constitute NIOSH endorsement of the sponsoring organizations or their programs or products. Furthermore, NIOSH is not responsible for the content of these Web sites. All web addresses in this document were accessible as of the date this manuscript was prepared for publication.

ORDERING INFORMATION

This document is in the public domain and may be freely copied or reprinted. To receive NIOSH documents or more information about occupational safety and health topics, contact NIOSH at

1–800–CDC–INFO (1–800–232–4636)
TTY: 1–888–232–6348
E-mail: cdcinfo@cdc.gov

or visit the NIOSH Web site at **www.cdc.gov/niosh**

For a monthly update on news at NIOSH, subscribe to
NIOSH eNews by visiting **www.cdc.gov/niosh/eNews**

DHHS (NIOSH) Publication Number 2009–100

November 2008

SAFER • HEALTHIER • PEOPLE™

Foreword

The United States currently depends on approximately 1.1 million fire fighters to protect its citizens and property from losses caused by fire. Each year in the United States, approximately 100 fire fighters die in the line of duty. Sudden cardiac death is the leading cause of fatalities, followed by trauma. In 1998, Congress appropriated funds to the National Institute for Occupational Safety and Health (NIOSH) for a fire fighter safety initiative. As part of this initiative, NIOSH developed and implemented the Fire Fighter Fatality Investigation and Prevention Program (FFFIPP).

The overall goal of the NIOSH FFFIPP is to reduce the number of fire fighter fatalities. To accomplish this goal, NIOSH conducts investigations of line-of-duty fire fighter deaths to identify contributing factors and to generate recommendations for prevention.

This document is a synthesis of the 1,286 individual recommendations from the 335 FFFIPP investigations conducted from 1998 to 2005. We hope that the fire service will use this document as a resource and catalyst for developing, updating, and implementing effective policies, programs, and training to prevent fatalities among fire fighters.

Christine M. Branche, Ph.D. /s
Acting Director
National Institute for Occupational
  Safety and Health
Centers for Disease Control
  and Prevention

Contents

Foreword iii

Acronyms vi

Executive Summary vii

Acknowledgments xi

I   **Recommendations for Preventing Fatalities Related to Medical Conditions** 1

      1. Medical screening 2
      2. Fitness and wellness 5

II   **Recommendations for Preventing Fatalities Related to Trauma** 9

      3. Standard operating procedures and guidelines (SOPs).... 10
      4. Communications 13
      5. Incident command 16
      6. Motor vehicle 19
      7. Personal protective equipment (PPE) 23
      8. Strategies and tactics 25
      9. Rapid intervention team 28
      10. Staffing 30

III **Conclusions** 33

IV **References and Additional Resources** 35

      References 36
      Additional NIOSH Fire Fighter Publications 39
      Questions? 41

Acronyms

CPR	cardiopulmonary resuscitation
FFFIPP	Fire Fighter Fatality Investigation and Prevention Program
IAFC	International Association of Fire Chiefs
IAFF	International Association of Fire Fighters
IC	incident commander
ICP	incident command post
ICS	incident command system
IDLH	immediately dangerous to life and health
METS	metabolic equivalents
MVA	motor vehicle accident
NIOSH	National Institute for Occupational Safety and Health
NFPA	National Fire Protection Association
NVFC	National Volunteer Fire Council
OSHA	Occupational Safety and Health Administration
PASS	personal alert safety system
PPE	personal protective equipment
RIT	rapid intervention team
SCBA	self-contained breathing apparatus
SOPs	standard operating procedures and guidelines
USFA	United States Fire Administration

Executive Summary

This document summarizes the most frequent recommendations from the first 8 years of the NIOSH Fire Fighter Fatality Investigation and Prevention Program (FFFIPP). The overall goal of the program is to reduce the number of fire fighter fatalities.

Through 2005, the FFFIPP investigated 335 fatal incidents involving 372 fire fighter fatalities. The investigations encompassed a variety of circumstances such as cardiovascular-related deaths, motor vehicle accidents, structure fires, diving incidents, and electrocutions. Fatalities have been investigated in career, volunteer, and combination departments in both urban and rural settings throughout the United States.

This document shares the most common recommendations from the 335 investigations and more than 1,286 recommendations that were developed by NIOSH investigators. These recommendations were developed using existing fire service standards, guidelines, standard operating procedures, and other relevant resources over the first eight years of the program. Fire departments can use this document when developing, updating, and implementing policies, programs, and training for fire fighter injury prevention efforts.

NIOSH Fire Fighter Fatality Investigation and Prevention Program (FFFIPP)

Each year, approximately 100 fire fighters die in the line of duty. Sudden cardiac death is the leading cause of on-duty fatalities, followed by trauma [USFA 2006]. In 1998, Congress appropriated funding to NIOSH for a fire fighter safety initiative. As part of this initiative, NIOSH implemented the FFFIPP to conduct independent, onsite investigations of fire fighter line-of-duty deaths. The goals of the FFFIPP are to

— better define the characteristics of line-of-duty deaths among fire fighters,

— develop recommendations for preventing deaths and injuries, and

— disseminate prevention strategies to the fire service.

The NIOSH investigations are voluntary, and names remain confidential. The purpose is not to find fault or cast blame on fire departments or individual fire fighters. During a fatality investigation, investigators evaluate the incident through a review of records such as police, medical, and victims' work/training records as well as departmental procedures. Investigators also examine the incident site and equipment used, including personal protective equipment.

Executive Summary

Interviews are conducted with members of the fire department, department investigators, and family members (when appropriate). Together, these sources build a picture of the circumstances of the event.

Next, the investigators review best practices, National Fire Protection Association (NFPA) standards, information from the United States Fire Administration (USFA), and the public health and fire service literature related to each case. They develop recommendations for prevention and write a narrative report describing the event and the recommendations. References for each recommendation can be found in the NIOSH Fatality Investigation Reports (Recommendations/Discussion section) on the NIOSH Web site (www.cdc.gov/niosh/fire/). Reports are disseminated to the fire service through a NIOSH Web site. Packets of selected reports have been periodically mailed to approximately 35,000 fire departments across the United States.

From 1998 through 2005, 863 fire fighters died in the line of duty (excluding the 343 fatalities in the World Trade Center tragedy) [USFA 2006]. The FFFIPP investigates both cardiovascular and traumatic incidents in career, volunteer, and combination departments. From 1998 to 2005, the FFFIPP investigated 335 of these fatal incidents, which accounted for 372 (43%) fire fighter fatalities. Therefore, the leading recommendations reflect the types of incidents investigated and may not apply to all deaths. Tables 1 and 2 list NIOSH fire fighter fatality investigations by cause of death and type of department for medically related and traumatic incidents.

Table 1. NIOSH investigations of medically related fire fighter fatalities by cause of death and type of department, 1998–2005

Cause of death	Type of department			Total
	Volunteer	Career	Combination	
Sudden cardiac death	37	69	37	143
Heat stress	0	2	0	2
Sarcoidosis—cardiac involvement	0	2	0	2
Other*	2	3	2	7
Total	39	76	39	154

*Other = pulmonary conditions, stroke, aneurysm, seizure disorder, and drug intoxication.

Table 2. NIOSH investigations of trauma-related fire fighter fatalities by cause of death and type of department, 1998–2005

Cause of death	Type of department			
	Volunteer	Career	Combination	Total
Asphyxiation	19	31	7	57
Electrocution	2	3	1	6
Burns	4	12	8	24
Drowning	1	6	1	8
Motor vehicle accident (MVA)	31	14	13	58
Non-MVA trauma	12	12	2	26
Other	1	1	0	2
Total	70	79	32	181

Leading Recommendations, 1998–2005

This document summarizes the leading recommendations from the 335 investigations conducted over the first 8 years of the NIOSH FFFIPP. The recommendations are grouped into 10 categories:

Recommendations for preventing fatalities related to medical conditions

1. Medical screening
2. Fitness and wellness

Recommendations for preventing fatalities related to trauma

3. Standard operating procedures and guidelines (SOPs)
4. Communications
5. Incident command
6. Motor vehicle
7. Personal protective equipment
8. Strategies and tactics
9. Rapid intervention team
10. Staffing

These recommendation categories are relevant to all types of departments in the fire service and should be considered when developing, updating, and implementing fire fighter injury prevention efforts.

How to Use This Document

Each of the 10 recommendation categories for preventing on-duty fire fighter fatalities includes the following sections:

- Overview of the category
- Category recommendations
- A case example of a fatality investigation report summary
- Fire department self-assessment questions
- Key resources

To illustrate each of the 10 recommendation categories (2 for fatalities related to medical conditions and 8 for fatalities related to trauma), a fatality investigation report summary was chosen as an example for each category. Most fatalities involve multiple factors, and recommendations for preventing a fatal incident may be included in more than one recommendation category.

Acknowledgments

The authors convey a special thanks to the Division of Safety Research Surveillance and Field Investigations Branch for their contribution and review of the document.

Contributors:

Richard W. Braddee, Team Leader, Fire Fighter Fatality Investigation and Prevention Program

Virginia Lutz, Safety and Occupational Health Specialist

Mark F. McFall, Safety and Occupational Health Specialist

Jay L. Tarley, Safety and Occupational Health Specialist

Anne C. Hamilton edited the document. Gino Fazio and Vanessa Becks provided desktop design and production.

The authors thank the following persons for their participation in the review of this technical document:

Nelson Bryner, National Institute of Standards and Technology, Leader Fire Fighting Technology Group, Fire Research Division

Daniel L. Gregory, Fire Department Safety Officers Association, Chairman

Heather Schaffer, National Volunteer Fire Council, Director

J. Gordon Routley, Fire Service Consultant

Robert E. Solomon, National Fire Protection Association, P.E., Assistant Vice President, Building and Life Safety Codes

Bruce Varner, International Association of Fire Chiefs, Chair Safety and Health Committee

The authors also thank Lunette Utter, Division of Safety Research, who provided database management.

I Recommendations for Preventing Fatalities Related to Medical Conditions

1. Medical screening 2
2. Fitness and wellness 5

The terms cardiovascular disease, heart attacks, coronary artery disease, and sudden cardiac death appear frequently in the next two recommendations categories. Heart attacks and coronary artery disease are two conditions under the umbrella term cardiovascular disease. Not all sudden cardiovascular events result in sudden death, typically defined as the unexpected natural death due to a cardiac cause within a short period of time after the onset of acute symptoms.

1. Medical screening

Sudden cardiac death is the leading cause of fire fighter fatalities, representing approximately 45% of all line-of-duty deaths [USFA 2006]. The FFFIPP conducted 154 fatality investigations related to medical conditions from 1998 to 2005. Most of these sudden cardiac deaths were related to heart attacks (also known as myocardial infarctions). Heart attacks occur when there is a blockage of coronary artery blood flow which can be triggered by heavy physical exertion probably due to acute increases in heart rates and blood pressure [Willich et al. 1993]. Typically, fire fighters have elevated heart rates and blood pressures when responding to alarms conducting fire supression tasks on the fireground or during physically demanding training excercises [Gledhill and Jamnik 1992].

A case-control study using the NIOSH on-duty cardiovascular fatality data on fire fighters showed that participation in fire suppression and training and response to an alarm were significantly associated with fatalities from coronary heart disease [Kales et al. 2003, 2007]. In addition, NIOSH found that most victims had multiple risk factors for coronary artery disease. Despite these findings, only 48 (31%) of the 154 departments where NIOSH investigated fatalities from cardiovascular disease conducted annual screening programs on all fire fighters to identify risk factors for coronary artery disease. Of the 48 departments conducting screening, only 16 (33%) conducted exercise stress tests on those at increased risk for coronary artery disease and sudden cardiac death [NIOSH 2007].

Recommendations for fire departments

- Conduct annual medical evaluations to screen all fire fighters for risk factors for coronary artery disease (e.g., smoking, diabetes mellitus, high blood pressure, high blood cholesterol, physical inactivity, obesity, and a family history of coronary artery disease).

- Conduct exercise stress tests on fire fighters who have coronary artery disease or who are at increased risk of this disease and sudden cardiac death. Increased risk for sudden cardiac death is defined as male fire fighters older than 45 years of age (older than 55 years for female fire fighters) with two or more risk factors for coronary artery disease (listed above) (Key Resource 2 on page 4).

- Ensure that fire fighters are medically cleared by physicians who are knowledgeable about the cardiovascular demands of fire fighting and aware of published medical guidelines for fire fighters.

Case 1. Fire fighter dies after collapse at apartment building fire

A 49-year-old male career captain with coronary artery disease experienced sudden cardiac death while serving as the incident commander (IC) at a structure fire. Before his collapse, the captain had not been performing physically demanding tasks. When crew members found him, he was unresponsive, without a pulse, and not breathing. Despite immediate cardiopulmonary resuscitation (CPR) and advanced life support, the captain died. No autopsy was performed.

Two years earlier, the captain had been diagnosed with coronary artery disease (100% occlusion of the right coronary artery). His treatment included a coronary artery stent, smoking cessation, and quarterly checkups with his private cardiologist. Follow-up studies by his private physician revealed a remote (old) myocardial infarction, a mildly dilated left ventricle, and a negative exercise (treadmill) stress test with fair exercise tolerance (10 metabolic equivalents [METS]). Two months before his death, the captain experienced a sudden loss of consciousness (syncope). A work-up by his private cardiologist revealed a slightly enlarged heart with slight left ventricle dilation, mildly impaired pumping function (decreased ejection fraction), and heart arrhythmia (atrial fibrillation with premature ventricular complexes). The captain was treated with Coumadin® (anticoagulant drug) and released for unrestricted duty.

This case illustrates the need for medical screening because the fire department did not conduct annual or periodic medical evaluations. The captain's medical evaluations and work clearances were conducted by his private physician. The captain had five medical conditions that had the potential to interfere with his essential job tasks: atrial fibrillation, anticoagulant drug use, cardiac enlargement, coronary artery disease, and syncope. If he had been examined by a provider familiar with consensus medical standards for fire fighters [NFPA 1582*], he might have been restricted from full fire suppression duties—which might have prevented his sudden cardiac death at this time [NIOSH 2002a].

Assess your department

1. Does our fire department physician know the status of member's risk factors for coronary artery disease?
2. Does our fire department require exercise stress tests for fire fighters at increased risk for coronary artery disease and sudden cardiac death?

---

*See Key Resources

3. Are members under the care of medical providers that are aware of fire service guidelines regarding medical clearance for duty?

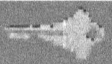 **KEY RESOURCES**

1. NFPA 1582: Standard on comprehensive occupational medical program for fire departments. Available at www.nfpa.org/catalog/search.asp?action=search&query=1582

2. ACC/AHA [2002]. Guidelines update for exercise testing: a report of the American College of Cardiology/American Heart Association Task Force on Practice Guidelines (Committee on Exercise Testing). Available at http://content.onlinejacc.org/cgi/content/full/40/8/1531

3. NIOSH [2007]. NIOSH Alert: preventing fire fighter fatalities due to heart attacks and other sudden cardiovascular events. Available at www.cdc.gov/niosh/docs/2007-133/

2. Fitness and wellness

Fitness and wellness programs can reduce modifiable coronary artery disease risk factors (Box 1). These programs are effective at reducing coronary artery disease risk factors, and can be cost effective, typically by reducing the number of work-related injuries and lost workdays in both fire fighters [Garfi et al. 1996; Harger et al. 1999; Dempsey et al. 2002; Stevens et al. 2002; Womack et al. 2005; Blevins et al. 2005; Kuehl 2007] and other workers [Maniscalco et al. 1999; Stein et al. 2000; Aldana 2001]. Programs that include individualized risk reduction for high-risk workers within the context of a comprehensive program seem to hold the most promise for positive clinical and cost outcome [Pelletier 2001]. The guidelines developed by the International Association of Fire Fighters (IAFF)/International Association of Fire Chiefs (IAFC), the National Volunteer Fire Council (NVFC)/United States Fire Administration (USFA), and the NFPA involve comprehensive programs with individualized assessment for all fire fighters (Boxes 2–4). Programs must be implemented in a positive, non-punitive manner.

Unfortunately, most of the fire departments in which NIOSH investigated an on-duty fatality from cardiovascular disease did not have a comprehensive fitness and wellness program as recommended by NFPA 1583, the IAFF/IAFC, and the NVFC/USFA documents (Boxes 2–4). From 1998 to 2005, 63 fire departments that experienced a fire fighter's death from cardiovascular disease, (41% of the 154 that NIOSH-investigated) had fitness programs, but only 16 of the 154 (10.4%) required participation.

Recommendations for fire departments
- Develop individualized fitness and wellness programs for all fire fighters.

Box 1

Modifiable risk factors for coronary artery disease

- High blood cholesterol
- Cigarette smoking
- High blood pressure
- Diabetes mellitus
- Obesity and overweight
- Physical inactivity

Source: American Heart Association www.americanheart.org/presenter.jhtml?identifier=4726

- Conduct annual fitness evaluations by a fitness coordinator under the supervision of a physician who is knowledgeable about the physical demands of fire fighting and aware of published medical guidelines for fire fighters (Boxes 2 and 3).
- Include health promotion components (e.g., smoking cessation, cancer screening, diet and nutritional education, and immunizations) (Boxes 2 and 3) in the fire department's wellness program.

Case 2. Fire fighter dies during live fire training

A 56-year-old career captain experienced sudden cardiac death during live-fire training. The live-fire exercise was in a two-story structure, and the training was in full compliance with all components of NFPA 1403, Standard on Live Fire

Box 2

The IAFF and IAFC Fire Service Joint Labor Management Wellness/Fitness Initiative

The IAFF/IAFC Wellness Initiative is an individualized program that requires the participation of all uniformed personnel in a nonpunitive manor. The Initiative includes a manual with information on the following topics:

- Fitness evaluation
- Medical evaluation
- Rehabilitation
- Behavioral health
- Data collection

The IAFF/IAFC Task Force has determined that successful implementation of the Wellness/Fitness Initiative requires a fire fighter in each department who can take the lead. This person must have the ability to design and implement fitness programs, improve the wellness and fitness of his or her department, and assist with the physical training of recruits. This need for a department-level leader led to the development of the Fire Service Peer Fitness Trainer certification program. The Peer Fitness Trainer program is being developed in conjunction with the American Council on Exercise.

For more information about this initiative, visit www.iaff.org/HS/Well/wellness.html

> **Box 3**
>
> **The NVFC Heart-Healthy Firefighter Program**
>
> The NVFC Heart-Healthy Firefighter Program promotes fitness, nutrition, and health awareness. The program includes
>
> - Heart healthy fire fighter kit.
> - Fired up for fitness challenge.
> - Heart healthy resource guide (Key Resouce 2 on page 8).
>
> Additional publications developed by the NVFC in conjunction with the USFA include the Health and Wellness Guide for the Volunteer Fire Service (www.usfa.dhs.gov/downloads/pdf/publications/fa-267.pdf) and the Emerging Health and Safety Issues in the Volunteer Fire Service (www.usfa.dhs.gov/downloads/pdf/publications/fa_317.pdf).

Training Evolutions. The captain, wearing full turnout gear and self-contained breathing apparatus (SCBA) with air on, had performed 8 evolutions/fire starts lasting 10 minutes each with only one 15-minute break. On his last evolution, he ignited the fire and exited from the structure in no apparent distress. Shortly thereafter, the captain solicited help from a paramedic in the stand-by ambulance. The paramedic's initial evaluation found a weak pulse and labored respirations. While removing his turnout gear, the captain collapsed. Despite immediate CPR and advanced cardiac life support, the captain died. The medical examiner who conducted the autopsy listed the cause of death as "probable cardiac arrhythmia due to ischemic heart disease due to severe coronary artery atherosclerosis."

The NIOSH investigation revealed that annual medical evaluations are required by this fire department. In fact, 2 months before his death, the captain was medically evaluated by a contract physician. The evaluation revealed high cholesterol, physical inactivity, obesity, and limited exercise tolerance. The contractor provided a voluntary wellness program that included nutrition and wellness education along with an individualized fitness plan. Unfortunately, the captain did not take advantage of this voluntary program [NIOSH 2002b].

Box 4

NFPA 1583, Standard on Health-Related Fitness Programs for Fire Fighters

NFPA 1583 stipulates that fire departments establish and provide a health-related fitness program that enables members to develop and maintain a level of health and fitness to safely perform their assigned functions. Components of this program should include the following:

- Educational program regarding description and benefits.
- Individualized exercise prescription with warm-up, aerobic, muscular, flexibility, healthy back, and cool-down guidelines.
- Safety and injury program.

For more information about this NFPA standard, visit www.nfpa.org/catalog/search.asp?action=search&query=1583

Assess your department

1. Does our fire department offer a fitness program and do members participate?
2. Do fire fighters receive a fitness evaluation every year?
3. Does our fire department offer a wellness program that includes help with smoking cessation and health screens, and do members participate?

KEY RESOURCES

1. NFPA 1583: Standard on health-related fitness programs for fire department members. Available at www.nfpa.org/catalog/search.asp?action=search&query=1583

2. NVFC/USFA health and wellness guide for volunteer fire service. Available at http://healthy-firefighter.org/

3. IAFF/IAFC fire service joint labor management wellness-fitness initiative. Available at www.iafc.org/publications/index.asp

4. NFFF National Fallen Firefighter Foundation. 16 Firefighter Life Safety Initiatives. Available at www.everyonegoeshome.org/initiatives.html

II Recommendations for Preventing Fatalities Related to Trauma

3. Standard operating procedures and guidelines (SOPs) 10
4. Communications 13
5. Incident command 16
6. Motor vehicle 19
7. Personal protective equipment (PPE) 23
8. Strategies and tactics 25
9. Rapid intervention team 28
10. Staffing 30

3. Standard operating procedures (SOPs) and guidelines

SOPs guide the actions of fire personnel in nearly all types of incidents. They describe expectations and responsibilities of team members. The development, periodic review, evaluation, and updating of SOPs are critical for an effective and safe department response. All fire department personnel, both career and volunteer, must be aware of, trained in, and compliant with the department's SOPs (Box 5) [USFA 1999a; Brunacini 1985; NFPA 2006]. The most common types of recommendations in this section concern (1) developing SOPs specifically for incident command, safe driving, and seat belt use and (2) enforcing those SOPs.

Recommendations for fire departments

- Ensure that the department's SOPs are developed and followed and that refresher training is provided.
- Develop and enforce SOPs for the safe and prudent operation of emergency vehicles.
- Enforce SOPs in the use of seat belts for all emergency vehicles.
- Establish and implement an incident command system (ICS) with written SOPs for all fire fighters.
- Ensure that SOPs addressing emergency scene operations such as basement fires are developed and followed on the fireground.
- Develop and implement written maintenance procedures for the SCBA.

Box 5

Elements of SOPs

SOPs should be

- Written
- Official
- Applied to all situations
- Enforced
- Integrated into the management system

Source: Brunacini [1985].

Case 3. Two volunteer fire fighters die fighting a basement fire

A 29-year-old male volunteer lieutenant and a 32-year-old male volunteer fire fighter died while fighting a basement fire in a single-family home. The size-up revealed that the structure was empty, that smoke had been seen in the first floor bathroom, and that light gray smoke was coming from the eaves and around the chimney. No flames were visible. The IC ordered roof ventilation while two other fire fighters searched for the location of the fire on the first floor of the structure. The fire fighters initially did not encounter any fire. But on the way out, part of the dining room ceiling collapsed, and fire began rolling out of the collapsed area. They extinguished the fire and reported to the IC that they thought the fire was in the attic. As a result of this information, a second hole in the roof was cut, but only light smoke escaped. The IC therefore ordered a crew to reenter the structure, where they found and extinguished fire under the sill plate in the wall adjacent to the basement stairs. This information led them to believe they had a working basement fire. The IC sent a crew to enter the basement to search for and extinguish the fire. They found the fire at the ceiling level in the basement and were able to push it back before they had to evacuate due to a crew member's low-air alarm sounding.

A second crew entered the basement and encountered light fire conditions that they knocked down before they exited. A third crew entered the basement with no visible fire. The crew used a thermal imaging camera to search for hot spots and pulled the ceiling in the basement to extinguish any fire they found. Suddenly, conditions changed. The captain leading the search reported that an intense heat had engulfed him, pierced through his hood, and knocked him to his knees. The entire basement, including the stairwell, had become engulfed in flames. He ordered his crew to get out. They did so by crawling up the stairs to exit as thick black smoke rolled out behind them. Not all members of the search crew escaped. Two fire fighters were trapped. After multiple attempts, they were located; but their SCBAs had run out of air.

The department involved in this incident had no SOPs. If SOPs for emergency operations on this fireground had been established, they would have provided guidance for the IC on how to approach this fire. SOPs for emergency operations should cover specific operations such as ventilation, water supplies, and basement fires (which present a complex set of circumstances).

To minimize the risk of serious injury to fire fighters, written SOPs should be developed, followed, and included in the overall risk management plan for the fire department. If the SOPs are changed, appropriate training should be provided to all affected members. In this case, NIOSH made recommendations for

SOPs as well as for incident command and rapid intervention teams [NIOSH 2001a].

Assess your department

1. Have we ever worked an incident without SOPs? How organized and in control were we during this incident?
2. Do all fire fighters in the department follow SOPs and how are they enforced?
3. Is our department trained in SOPs?
4. Where do we keep our SOPs?
5. When was the last time our SOPs were reviewed and updated?

KEY RESOURCES

1. The USFA text, Developing Effective SOP, provides a complete guide (including templates) for the development of SOPs, 1999. Available at www.usfa.fema.gov/downloads/pdf/publications/fa-197.pdf

2. Firehouse.com has free downloadable samples of SOPs:

   www.firehouse.com/tech/sop.html

   www.firehouse.com/links/Fire_and_EMS_Department_SOPs/General/

   www.firehouse.com/links/Fire_and_EMS_Department_SOPs/

   www.firehouse.com/links/Fire_and_EMS_Department_SOPs/Hazardous_Materials_Team/

3. Firehouse.com [2001]. SOP: instructors guide Maryland Fire and Rescue Institute. A teaching guide for fire fighter trainers on basic knowledge of SOPs importance and implementation. Available at www.firehouse.com/training/drills/files/DM_0110.pdf

4. Cook JL [1998]. Standard operating procedure and guidelines. Saddle Brook, NJ: Penn Well. A textbook on designing appropriate SOPs.

5. NFPA 1500: Standard on fire department occupational safety and health program. Available at www.nfpa.org

6. NFPA 1561: Standard on emergency services incident management system. Available at www.nfpa.org

4. Communications

Communications within a fire department include elements that allow the exchange of information from different sources (Box 6). This flow of information leads to timely actions that allow the department to respond appropriately to calls [USFA 1995].

Problems with communications can be divided into two broad areas: mechanical/technical issues and human factors [USFA 2003a]. The recommendations in this category address both areas. For example, mechanical/technical issues as well as human factors are involved when fire personnel must use two-way radios, send emergency/evacuation signals, or report interior conditions to the IC.

Box 6

How well does your communication system work?

Analyzing a fire department's communications performance is important. USFA recommends that each fire department develop criteria to define acceptable performance for each aspect of the communications system.

The categories of criteria are as follows:

1. Acceptable time frame from the time the call is received to the time it is acknowledged by responding units
2. Adequate communication channels for tactical operations
3. The requirement for using communication protocols
4. The communications requirements for multiple incidents
5. An emergency backup plan for the communications system if the component systems fail
6. A complete tracking of all resources

Source: USFA [1995]. Fire department communications manual: a basic guide to system concepts and equipment. Emmitsburg: Department of Homeland Security, Federal Emergency Management Agency, United States Fire Administration, Section 3–16.

Recommendations for fire departments

- Establish a method of fireground communication that permits coordination between the incident commander and the fire fighters.
- Ensure that fire fighters are equipped with radios that do not bleed over, cause interference, or lose communication under field conditions.
- Consider providing all fire fighters with portable radios or radios integrated into their facepieces.
- Ensure that a tone or alert recognized by all fire fighters is transmitted immediately when conditions become unsafe for fire fighters.
- Instruct and train fire fighters in initiating emergency traffic (mayday-mayday) and activating their personal alert safety system (PASS) device when they become lost, disoriented, or trapped.
- Whenever a change in personnel occurs, make sure that everyone is briefed and understands the procedures and operations required for that shift, station, or duty.
- Ensure that properly functioning communications equipment is available and can adequately support the volume of radio traffic at fire scenes.
- Establish and maintain regional mutual-aid radio channels to coordinate and communicate activities involving units from multiple jurisdictions.

Recommendations for fire departments with diving rescue teams

- Ensure that positive communication is established among all divers and the personnel who remain on the surface.
- Ensure that divers maintain continuous visual, verbal, or physical contact with their dive partners.

Case 4. Structural collapse at residential fire claims lives of two volunteer fire chiefs and one career fire fighter

Three veteran fire fighters lost their lives in the collapse of a residential building. Multiple mutual aid departments were fighting the fire. The victims (two male volunteer chief officers and a male career fire fighter) were crushed in the interior of the building.

A primary search was being conducted and master streams were in operation when a fire fighter advised the IC that the second floor was about to give way. Another fire fighter from the second floor notified the IC that they were evacuating the second floor and the ceiling was coming down. The IC ordered the master streams to be shut down, but he did not call for a complete emergency evacuation for approximately 3 minutes.

A fire fighter from a mutual aid company attempted to radio his officer because he couldn't communicate with the IC and inform him that the floor was collapsing. The mutual aid fire fighter was broadcasting over a different frequency from the department in charge of the scene. The evacuation signal (air horns) was sounded and crews evacuated. The three victims and four other fire fighters reentered the structure's first floor to search for a missing fire fighter when the building totally collapsed. The victims and the four other fire fighters did not know that the missing fire fighter had been accounted for.

It is critically important to establish communication protocols with mutual aid companies and train together to enhance coordination and effectiveness during complex mutual aid operations. In addition to making recommendations about communications, NIOSH made recommendations about incident command and rapid intervention teams [NIOSH 2002c].

Assess your department

1. Have we ever experienced radio interference and bleed over? What have we done to correct this?

2. Does our department have communication SOPs with mutual aid companies?

3. Does our department have SOPs regarding fireground radio traffic, such as interior reports, size-up, completion of assignment, mayday, and evacuation signals?

 **KEY RESOURCES**

1. USFA [1995]. Fire department communications manual: a basic guide to system concepts and equipment. Available at www.usfa.fema.gov/downloads/pdf/publications/fa-160.pdf

2. USFA [2003]. Special Report: improving fire fighter communications. This report investigates potential causes of communications breakdown and provides recommendations that will help fire departments improve their operational communication. Available at www.usfa.dhs.gov/downloads/pdf/publications/tr-099.pdf

3. NFPA 1221: Standard for the installation, maintenance, and use of emergency services communication systems. Available at www.nfpa.org/catalog/search.asp?action=search&query=nfpa+1221

5. Incident command

The IC is responsible for the overall management of the incident. The IC's priorities are life safety, incident stabilization, and property conservation [USFA 1999b]. The incident command category includes implementation, size-up, determination of risk versus gain, and accountability of fire fighters.

Recommendations for fire departments

- Clearly identify the IC as the only person responsible for the overall coordination and direction of all activities at an incident.
- Ensure that the IC maintains the role of director and does not become involved in fire-fighting operations.
- Implement the ICS for the management of all fires and establish an incident command post (ICP) as needed to facilitate command and control, especially on complex fires involving multiple agencies.
- Appoint a separate incident safety officer (independent from the IC).
- Ensure that the IC conducts a complete size-up of the incident before initiating fire-fighting efforts and continually evaluates the risk versus gain during operations.
- Convey strategic decisions through the IC to all suppression crews on the fireground.
- Maintain accountability for all personnel at the fire scene.
- Train fire fighters to communicate interior conditions to the IC as soon as possible and to provide regular updates.

Case 5. First-floor collapse during residential basement fire claims lives of two fire fighters (career and volunteer) and injures a career fire fighter captain

A 28-year-old male volunteer fire fighter and a 41-year-old male career fire fighter died after becoming trapped in a house fire. The victims were from a mutual aid company. As they manned a hose and entered the structure, the floor collapsed. Both victims were trapped in the basement.

Although this incident had an IC in place, some operations were directed by other personnel and some operations were not in line with the tactics of the IC. An effective fireground operation revolves around one IC. Three recommendations were made for this case in the following areas: (1) the IC should be clearly identified as the only person responsible for the overall coordination and direction of all activities at an incident; (2) the IC should convey strategic decisions to all suppression crews on the fireground and continually reevaluate the fire condition; and (3) the IC should conduct an initial size-up of the incident before initiating fire-fighting efforts and should continually evaluate the risk versus gain during operations at an incident. In addition to making recommendations related to incident command, NIOSH also made strategy and tactics recommendations [NIOSH 2002d].

Assess your department

1. Does our department implement ICS at each incident—large and small?

2. What training does our department provide in ICS? Is this training sufficient?

3. What accountability system is used in our department? Does the IC know the location of company members at all times?

4. Do we evaluate risk versus gain before initiating each offensive, interior attack and continually throughout an incident?

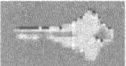
KEY RESOURCES

1. Occupational Safety and Health Administration (OSHA) ICS eTools are illustrated, interactive Web-based training tools on occupational safety and health topics. As indicated in the disclaimer, eTools do not create new OSHA requirements. This eTool is designed to provide basic information about the Incident Command System and the Unified Command, specifically as it relates to the National Contingency Plan 40 CFR.300. Available at www.osha.gov/SLTC/etools/ics/index.html

2. USFA Incident Command and Control Simulation Series (CD-ROM): This program provides the user with a basic explanation of the ICS, including organizational structure, position and functions, responsibilities and considerations. Available at www.usfa.fema.gov/applications/publications/browse.cfm?sc=18

3. Brunacini [2002]. Fire command text, workbook, and videos. Available at www.nfpa.org/catalog/search.asp?action=search&query=Fire+Command&x=16&y=192

4. USFA [1996]. Personnel accountability system technology assessment. Report detailing how to implement effective communication between IC and fire fighters. Available at www.usfa.fema.gov/downloads/pdf/publications/fa-198.pdf

5. NFPA 1561: Standard on emergency services incident management system. Covers minimum requirements for an incident management system to be used by fire departments to manage all emergency incidents. Available at www.nfpa.org/catalog/search.asp?action=search&query=nfpa+1561

6. Motor vehicle

Recommendations concerning motor vehicles are defined as issues related to vehicle operators and passengers in all types of vehicles (e.g., tankers, privately owned vehicles). This category includes recommendations such as prudent driving, seat belt use (Figure 1), maintenance, and operator training (Box 7).

Box 7

Driver training at the Sacramento Regional Training Facility

The Sacramento Regional Training Facility [www.srdtf.com/] has a driver training course with the following characteristics:

Primary course

- 8 hours of defensive driving
- 16 hours of actual course driving
- 4 hours in the simulator
- Focus on vehicle placement and skid recovery during the driving portion of the course

Instructors

- All are Fire Course Instructors and Peace Officer Standards and Training-certified.
- All have completed 40-hour course on law and liability issues.
- Most (75%) are certified to conduct driver motor vehicle training and examinations.
- All are cross-trained, permitting them to train police personnel.

Results

- From 1999 to 2003, fire department at-fault accidents decreased by 77%.
- Each year since implementation of the training, an estimated $2 to $4 million is saved in litigation cost.

Source: USFA [2004]. Emergency vehicle safety initiative. Emmitsburg, MD: Department of Homeland Security, Federal Emergency Management Agency, U.S. Fire Administration [www.usfa.fema.gov/downloads/pdf/publications/fa-272.pdf].

Seat belt use

Seat belts are the most effective means of reducing injuries in motor vehicle crashes [Dinh-Zarr et al. 2001]. However, fire fighters use seat belts less often than the general population. (See Figure 1).

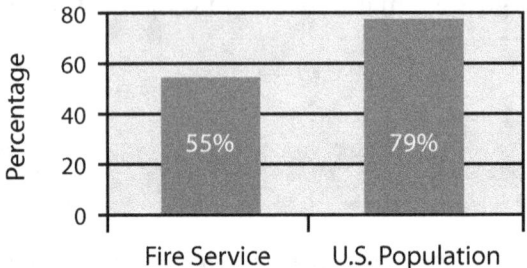

Figure 1. Comparison of seat belt use for fire service personnel and U.S. population. Sources: Firehouse.com (www.firehouse.com/polls/2003/), National Highway Traffic Safety Administration (www.nhtsa.dot.gov/portal/site/nhtsa/menuitem.f2217bee37fb 302f6d7c121 046108a0c/;jsessionid=EJkeuZc jL2N3I95 QImc8jCou5J4yJ7hkHsZC t3OFRVOkucvrD3ei! -626933413).

Recommendations for fire departments

- Ensure that all fire apparatus are equipped with seat belts.
- Ensure that all fire fighters riding in emergency fire apparatus are wearing seat belts and are belted securely.
- Do not permit drivers of fire apparatus to move vehicles until all occupants are secured with seat belts.
- Inform all drivers of fire department vehicles that they are responsible for the safe and prudent operation of the vehicle under all conditions.
- Instruct drivers of emergency fire apparatus to come to a complete stop at intersections having a stop sign or a red signal light before proceeding through the intersection.
- Instruct drivers of fire department vehicles to come to a complete stop at all unguarded railroad grade crossings during emergency response or non-emergency travel.
- Provide drivers of fire department vehicles with driver training at least twice a year.
- Develop comprehensive apparatus maintenance programs in accordance with manufacturer's specifications and instructions. Make sure these include regularly scheduled inspections, documentation, and

procedures for removing apparatus from service until major defects are repaired.

- Provide baffles for all apparatus equipped with water tanks to control water movement.
- Determine a safe operating weight for water tankers based on vehicle characteristics, and remove overweight vehicles from service.
- Make sure that the placement of additional equipment (e.g., radios and map card boxes) on an apparatus does not interfere with the driver's ability to operate controls.

Case 6. Volunteer assistant chief dies in tanker rollover

A 46-year-old male volunteer assistant chief was fatally injured after being ejected from a water tanker as a result of a rollover crash. The victim was traveling to a wildland fire on an unpaved road. The tanker failed to negotiate a curve, rolled over, left the road, and rolled several more times down into a canyon. The victim was ejected from the cab during the rollover and was found lying unresponsive on the ground. He was pronounced dead at the scene. In the post-crash investigation, the State police reported that the tanker "probably had brake failure prior to the accident" and noted that the master cylinder was leaking brake fluid and the emergency brake was inoperable. The vehicle was not equipped with seat belts.

The department involved in this incident performs visual checks of their vehicles; however, no SOPs were in place for determining when to declare a vehicle unsafe and to remove it from service. Fire departments should ensure that all fire apparatus are equipped with seat belts [NIOSH 2003].

Assess your department

1. Are our vehicles equipped with seat belts?
2. Do fire fighters buckle up when traveling to and from all incidents?
3. Does our department have an SOP for ensuring seat belt use?
4. Are the apparatus drivers in the department trained, licensed, and careful operators?
5. Does our department maintain emergency vehicles properly? Do we have SOPs to remove unsafe equipment?
6. What is our department's SOP for training drivers?

 **KEY RESOURCES**

1. USFA. Emergency vehicle safety initiative. Available at www.usfa.fema.gov/downloads/pdf/publications/fa-272.pdf

2. USFA [1996]. Emergency vehicle driver training package. This package is designed to assist fire and emergency medical service (EMS) departments with training in emergency vehicle driving operations. Available at www.usfa.fema.gov/downloads/pdf/publications/fa-110.pdf

3. USFA [2003]. Safe operation of fire tankers. Available at www.usfa.fema.gov/downloads/pdf/publications/fa-248.pdf

4. NFPA 1451: Standard for fire service vehicle operations training program. Contains the minimum requirements for a fire service vehicle risk management program. Available at www.nfpa.org

5. NFPA 1002: Standard for fire apparatus driver/operator professional qualifications. Identifies the professional levels of competence required of the fire apparatus driver/operator. Available at www.nfpa.org

6. NFPA 1911: Standard for the inspection, maintenance, testing, and retirement of in-service automotive fire apparatus. Available at www.nfpa.org

7. NIOSH [2001]. Hazard ID: fire fighter deaths from tanker truck rollovers. Available at www.cdc.gov/niosh/hid14.html

8. USFA. Emergency vehicle safety program. Available at www.usfa.dhs.gov/research/safety/vehicle.shtm

9. IAFC. Policies and procedures for emergency vehicle safety. Available at www.iafc.org/displaycommon.cfm?an=1&subarticlenbr=602

10. NFFF National Fallen Firefighter Foundation. 16 Firefighter Life Safety Initiatives. Available at www.everyonegoeshome.org/initiatives.html

7. Personal protective equipment (PPE)

The use of appropriate PPE is critical for preventing fire fighter fatalities and injuries. The PPE recommendations address the proper use of PASS and SCBA devices and PPE for water incidents.

Recommendations for fire departments

- Properly inspect, use, and maintain SCBAs to ensure they function properly when needed.
- Ensure that officers enforce the use of and that fire fighters wear their SCBAs equipped with integrated PASS (including the initial assessment) whenever they might be exposed to a toxic or oxygen-deficient atmosphere.
- Ensure that fire fighters wear and use PASS devices when involved in fire fighting, rescue, and other hazardous duties.
- Ensure that personnel wear PPE suitable to the incident while operating at an emergency scene (e.g., a highly visible [strong yellow-green or orange] reflectorized flagger vest).
- Ensure that adequate PPE (e.g., SCBA) is available while fire fighters are engaged in fire activity.
- Provide all rescue personnel with appropriate PPE (i.e., water rescue helmet and an appropriate personal flotation device) when operating at a water incident and ensure its proper use.

Case 7. Volunteer fire fighter dies after being struck by a motor vehicle while directing traffic

A 48-year-old volunteer fire fighter was struck and killed by a vehicle while on the scene of a motor vehicle crash. He was thrown more than 32 feet and pinned under a stopped pickup truck. After extrication, he was transported to a local trauma hospital where he died the next day. At the time, he was wearing dark-colored street clothes while directing traffic in low light during the early evening.

He was not using a flashlight, red wand, or a reflective vest—all of which were available from the staged emergency apparatus. In addition, the rescue unit on scene had available traffic cones, flares, and turnout coats. Fire departments should ensure that personnel wear personal protective clothing suitable to that incident while operating at an emergency scene—for example, a highly visible reflectorized flagger vest (strong yellow-green or orange). Personnel need to be easily seen while directing traffic near an incident scene. In addition to recommendations related to personal protective equipment, NIOSH also made recommendations regarding SOPs at motor vehicle events [NIOSH 2001b].

Assess your department
1. What PPE does our department have for responding to roadway incidents?
2. How does our department ensure that all fire fighters wear and use their PASS appropriately?
3. Does our department have a SCBA maintenance program?

KEY RESOURCES

1. NFPA 1851: Standard on selection, care, and maintenance of protective ensembles for structural fire fighting and proximity fire fighting. This text specifies requirements for the selection, care, and maintenance of fire-fighting protective clothing including coats, trousers, hoods, helmets, gloves, and footwear. Available at www.nfpa.org

2. USFA [1992]. Minimum standards on structural firefighting protective clothing and equipment. A guide for fire service education and for PPE procurement. Available at www.usfa.fema.gov/downloads/pdf/publications/fa-137.pdf

3. USFA. Emergency vehicle safety initiative. (Discusses PPE for highway operations visibility.) Available at www.usfa.fema.gov/downloads/pdf/publications/fa-272.pdf

4. NFPA 1982: Standard on personal alert safety systems (PASS). Covers minimum performance criteria, functioning, and test methods for personal alert safety systems to be used by fire fighters engaged in rescue, fire fighting, and other hazardous duties. Available at www.nfpa.org/catalog/product.asp?category%5Fname=&pid=198207&target%5Fpid=198207&src%5Fpid=&link%5Ftype=search

8. Strategies and tactics

This category addresses common fireground strategies such as switching from offensive to defensive attack, evaluating risk versus gain, and using fire tactics such as ventilation, water supply and suppression, exit routes, safety line use, and searches (Box 8).

Box 8

How to evaluate effectiveness of your tactical operations

The USFA suggests four questions for ICs directing fireground operations:

1. Is my plan of managing this operation effectively working?
2. Do I need more resources?
3. Can I release any resources?
4. Are there any emergency or hazardous conditions that prevent the completion of my assignments?

Source: USFA [1999]. Managing company tactical operations, p. SM 1–25.

Recommendations for fire departments

- Ensure that fire-fighting tactics and operations do not increase hazards on the interior (e.g., opposing hose streams).
- Ensure that fire fighters from the ventilation crew and the attack crew coordinate their efforts.
- Evacuate fire fighters performing fire-fighting operations under or above trusses as soon as it is determined that the trusses are exposed to fire.
- Establish a collapse zone and clearly mark it at structure fires where buildings have been identified at risk of collapsing.
- Monitor the collapse zone to ensure that no fire-fighting operations take place in the danger zone.
- Ensure that fire fighters, when operating on the floor above the fire, have a charged hoseline.

- Ensure that any hose line taken into the structure remains inside until all crews have exited.
- Ensure that the IC conducts a complete size-up of the incident before initiating fire-fighting efforts and continually evaluates the risk versus gain during operations.
- Ensure that a backup line is manned and in position to protect exit routes.
- Ensure that backup lines are equal to or larger than the initial attack lines.
- Consider using a thermal imaging camera as a part of the exterior size-up.

Case 8. Volunteer lieutenant dies following structure collapse at a residential house fire

A 36-year-old volunteer lieutenant died after being crushed by an exterior wall that collapsed during a residential structure fire. The lieutenant was performing manual suppression activities several feet from an exterior wall that was weakened by the fire. The exterior wall collapsed onto the victim. He was freed from the debris within minutes, but attempts to revive him were unsuccessful and he was pronounced dead at the scene.

The IC must consider, on arrival and throughout the incident, whether the operation is to be conducted in an offensive or defensive mode. Fireground conditions had indicated questionable structural integrity; but defensive operations were never called, and a collapse zone was not established. A collapse zone must be established on arrival and at any time during a structure fire if size-up determines that structural integrity is questionable. A collapse zone is an area around and away from a structure in which debris might land if the structure fails. This area should be equal to the height of the building plus an additional allowance for debris scatter. At a minimum it should be equal to 1½-times the height of the building. For example, if a wall is 20 feet high, the collapse zone boundary should be established at least 30 feet away from the wall. In addition to recommendations related to strategies and tactics, NIOSH made recommendations regarding incident command and SOPs [NIOSH 2002e].

Assess your department

1. How does our department ensure that tactical operations do not increase hazards to fire fighters?
2. When entering a hazardous structure, do our team members always lay a hose line, rope, or some other type of exit guide?

3. Can all fire fighters in our department identify a truss roof or floor system and understand its collapse characteristics?

4. Do we evaluate risk versus gain each time before initiating an offensive interior attack and continually throughout an incident?

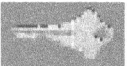
KEY RESOURCES

1. USFA [1999]. Managing company tactical operations: tactics. This textbook comprehensively covers important strategies and tactics such as ventilation, rescue, fire containment, water supply, and salvage. Available at www.usfa.dhs.gov/training/nfa/handoff/

2. NFPA 1710: Standard for the organization and deployment of fire suppression operations, emergency medical operations, and special operations to the public by career fire departments. Available at www.nfpa.org/aboutthecodes/Aboutthecodes.asp?DocNum=1710

3. NFPA 1720: Standard for the organization and deployment of fire suppression operations, emergency medical operations, and special operations to the public by volunteer fire departments. Available at www.nfpa.org/aboutthecodes/Aboutthecodes.asp?DocNum=1720

4. NIOSH [1999]. NIOSH Alert: Preventing injuries and deaths of fire fighters due to structural collapse. Available at www.cdc.gov/niosh/99-146.html

5. NFFF National Fallen Firefighter Foundation. 16 Firefighter Life Safety Initiatives. Available at www.everyonegoeshome.org/initiatives.html

9. Rapid intervention team

The USFA defines a rapid intervention team (RIT) as a "dedicated crew of fire fighters that is assigned for rapid deployment to rescue lost or trapped members" (Box 9) [USFA 2003b]. In some of the FFFIPP investigations, RITs were never assigned or did not remain available at an area designated by the IC.

Recommendations for fire departments

- Ensure that an RIT is established when fire fighters enter an immediately dangerous to life and health (IDLH) atmosphere and that the RIT is properly trained and equipped.
- Ensure that once an RIT is established, they remain the RIT throughout the operation.
- Ensure that only the assigned RIT completes search and rescue operations.

Box 9

RIT Equipment

The USFA suggests the following minimum equipment for an RIT:

1. Extra SCBA complete with harness, regulator, and extra masks (consider that mutual aid companies may use different SCBA systems)
2. Search rope
3. Forcible entry hand tools such as axe, sledge, halligan bar, and bolt cutters
4. Mechanical forcible entry tools such as chain saw, metal cutting saw, and masonry cutting saw
5. Hose line
6. Ladder complement
7. Thermal imaging camera
8. High intensity hand light

Source: USFA [2003]. Rapid intervention teams and how to avoid needing them, p. 36.

Case 9. Structural collapse at an auto parts store fire claims the lives of one career lieutenant and two volunteer fire fighters

Three veteran fire fighters (a career lieutenant and two volunteer fire fighters) were conducting an interior fire attack at an auto parts store when the ceiling collapsed, killing all three fire fighters. The IC had called for an evacuation of the building only moments before the ceiling collapsed. An RIT was established after the collapse. After multiple attempts by the RIT and other fire fighters, all three victims were recovered, but they were pronounced dead at the scene.

In this incident, an RIT should have been established, in position, and ready for deployment during the initial stages of an incident. Instead, an RIT was not established until the ventilation crew exited from the roof, changed out their air bottles, and became the RIT.

An RIT should respond to every major fire. The team should report to the IC and remain at the designated area until an intervention is required to rescue a fire fighter. The RIT should have all the tools necessary to complete the job—e.g., a search rope, rescue rope, first-aid kit, and resuscitator to use in case a fire fighter needs help. These teams can intervene quickly to rescue fire fighters who run out of breathing air or who become disoriented, lost in smoke-filled environments, trapped by fire, or involved in structural collapse. In addition to recommendations related to RITs, NIOSH made recommendations regarding incident command and PPE [NIOSH 2002f].

Assess your department

1. Does our department always implement an RIT?
2. Have we been caught without an RIT ready when we needed one?
3. Have we had any near misses?

KEY RESOURCES

1. USFA [2003]. RITs: how to avoid needing them. This USFA special report covers the tools to develop and improve RITs with examples from around the fire service. Available at www.usfa.fema.gov/downloads/pdf/publications/tr-123.pdf

2. NFPA [2001]. Rapid intervention teams. Comprehensive text on RITs. Available at www.nfpa.org

10. Staffing

This category addresses issues of adequate numbers of fire fighters, adequate deployment procedures, and team continuity.

Recommendations for fire departments

- Ensure that at least four fire fighters are on the scene before initiating interior fire-fighting operations at a structural fire: two-in and two-out.
- Ensure that adequate fire control forces and fire suppression equipment are on the scene and available for deployment for fire control activities.
- Ensure that adequate staff are available to immediately respond to emergency incidents.
- Maintain team continuity.

Case 10. Career fire fighter dies after single-family house fire

A 22-year-old male career fire fighter sustained fatal injuries while fighting a single-family house fire. The structure was fully involved as the first two responding fire fighters arrived on the scene, and a neighboring second house was 50% involved in the attic area. Mutual aid was called immediately. The victim arrived on a mutual aid tanker and was assigned by the IC to don his gear and back up the first two fire fighters. The victim approached one of the fire fighters first on the scene to offer help but instead was cautioned to be careful around the garage area because of heavy fire in the room above the garage. The fire fighter apparently was unaware that the IC had ordered the victim to act as backup. They did not coordinate efforts; instead, the first-on-scene fire fighter proceeded toward the structure to complete suppression activities and the victim went to the garage. Approximately 5 minutes later, the IC was notified by a bystander that the victim was lying halfway out of the garage area. Unknown to the IC, the victim had proceeded to the garage area alone. Shortly thereafter, a partial roof and garage door collapsed, trapping him. The IC ran to the garage area and helped pull the victim out from the debris. Emergency medical personnel moved the victim to the street and began administering first aid. He was flown to a regional hospital where he died 26 days later from his injuries.

The investigators noted that only two fire fighters were used to initiate an interior attack and that the procedure of two-in and two-out was not maintained. NFPA recommends having four persons (two-in and two-out), each with protective clothing and respiratory protection, as the minimum number essential for the safety of those performing work inside a structure. The team members should communicate with each other through visual, audible, or electronic means to coordinate all activities and to determine whether emergency rescue is needed. Also, the OSHA standard [29 CFR 1910.134] states that when at least two enter a structural fire-fighting atmosphere (IDLH), two must remain on the outside and maintain visual or voice contact to assist in emergency rescue activities. In addition to recommendations related to staffing, NIOSH made recommendations regarding incident command as well as strategies and tactics [NIOSH 2001c].

Assess your department

1. How does our department ensure and enforce two-in and two-out?

2. How well does our department coordinate with mutual aid companies? Do we work together or are there conflicts? How does this affect outcomes? How can we improve coordination?

3. Do we have sufficient staffing and equipment on the scene necessary to begin fire-fighting operations based on the expected fire conditions?

 **KEY RESOURCES**

1. NFPA 1710: Standard for the organization and deployment of fire suppression operations, emergency medical operations, and special operations to the public by career fire departments. Contains minimum requirements relating to the organization and deployment of fire suppression and emergency medical operations. Available at www.nfpa.org

2. NFPA 1720: Standard for the organization and deployment of fire suppression operations, emergency medical operations, and special operations to the public by volunteer fire departments. Contains minimum requirements relating to the organization and deployment of fire suppression resources (and for those fire departments that provide them, emergency medical and special operations resources). Available at www.nfpa.org

3. NFPA 1500: Standard on fire department occupational safety and health program. Available at www.nfpa.org

III Conclusions

This document summarizes the first 8 years of leading recommendations from the FFFIPP to support the fire service in the prevention of fire fighter deaths. All recommendations are pertinent to every department, and it is important for departments to implement these recommendations in their day-to-day operations, including training and SOPs.

In conclusion, the recommendations regarding cardiovascular health, wellness and fitness, SOPs, communications, incident command, motor vehicles, PPE, strategies and tactics, RIT, and staffing were consistently recommended by the NIOSH FFFIPP. To attain the goal of preventing fire fighter fatalities, these recommendations must be consistently implemented.

References and Additional Resources

References 36

Additional NIOSH Fire Fighter Publications 39

Questions? 41

References

Aldana SG [2001]. Financial impact of health promotion programs: a comprehensive review of the literature. Am J Health Promot 15:296–320.

Blevins JS, Bounds RG, Mougin JE, Coast JR [2005]. Minor changes in lifestyle programming significantly improve health and fitness in fire fighters. Med Sci Sports Exercise 37(5):S1–S405.

Brunacini AV [1985]. Fire command. Quincy, MA: National Fire Protection Association, pp. 15–16.

Dempsey WL, Stevens SR, Snell CR [2002]. Changes in physical performance and medical measures following a mandatory fire fighter wellness program. Med Sci Sports Exercise 34(5):S1–S258.

Dinh-Zarr T, Sleet D, Shults R, Zara S, Elder R, Nichols J, Thompson R, Sosin D, Task Force on Community Preventive Services [2001]. Reviews of evidence regarding interventions to increase the use of safety belts. Am J Prev Med 21:48–65.

Firehouse [2003]. 2003 Poll. www.firehouse.com/polls/2003/ Date accessed: July 23, 2007.

Garfi J, Marcotte J, Drury D, Ritterhaus C, Headley S [1996]. The effects of 16 weeks of cross training on resting blood pressure in firefighter recruits. Med Sci Sports Exercise 28(5):S1–S14.

Gledhill N, Jamnik VK [1992]. Characterization of the physical demands of firefighting. Can J Sports Sci 17(3):207–213.

Harger NB, Matthews MD, Kirk EP [1999]. Moderate aerobic work and regular resistive exercises improve selected fitness components in professional firefighters. Med Science in Sports and Exercise 31(5):S1–S376.

Kales SN, Soteriades ES, Christoudias SG, Christiani DC [2003]. Firefighters and on-duty deaths from coronary heart disease: a case control study. Environ Health: a global access science source. Vol. 2 [www.ehjournal.net/content/2/1/14].

Kales SN, Soteriades ES, Christophi CA, Christiani DC [2007]. Emergency duties and deaths from heart disease among firefighters in the United States. N Engl J Med 356:1207–1215.

Kuehl K [2007]. Economic Impact of the Wellness Fitness Initiative. Presentation at the 2007 John P. Redmond Symposium in Chicago, IL on October 23, 2007.

Maniscalco P, Lane R, Welke M, Mitchell J, Husting L [1999]. Decreased rate of back injuries through a wellness program for offshore petroleum employees. J Occup Environ Med 41:813–820.

NFPA [2006]. Codes and standards. Quincy, MA: National Fire Protection Association [www.nfpa.org/aboutthecodes/list_of_codes_and_standards.asp].

NHTSA [2003]. New Department of Transportation data show rising safety belt use rates in most States. Washington, DC: Department of Transportation, National Highway Traffic Safety Administration [www.nhtsa.dot.gov/portal/site/nhtsa/menuitem.302f6d7c121046108a0c/;jsessionid=EJkeuZcjL2N3I95QImc8jCou5J4yJ7hkHsZCt3OFRVOkucvrD3ei!- 626933413].

NIOSH [2001a]. Two volunteer fire fighters die fighting a basement fire—Illinois. Morgantown, WV: U.S. Department of Health and Human Services, Centers for Disease Control and Prevention, National Institute for Occupational Safety and Health, Fire Fighter Fatality Investigation and Prevention Program Report No. F2001–08 [www.cdc.gov/niosh/fire/reports/face200108.html].

NIOSH [2001b]. A volunteer fire fighter died after being struck by a motor vehicle while directing traffic—New York. Morgantown, WV: U.S. Department of Health and Human Services, Centers for Disease Control and Prevention, National Institute for Occupational Safety and Health, Fire Fighter Fatality Investigation and Prevention Program Report No. F2001–07 [www.cdc.gov/niosh/fire/reports/face200107.html].

NIOSH [2001c]. Career fire fighter dies after single-family-residence house fire—South Carolina. Morgantown, WV: U.S. Department of Health and Human Services, Centers for Disease Control and Prevention, National Institute for Occupational Safety and Health, Fire Fighter Fatality Investigation and Prevention Program Report No. F2001–27 [www.cdc.gov/niosh/fire/reports/face200127.html].

NIOSH [2002a]. Fire fighter dies after collapse at apartment building fire—Kentucky Morgantown, WV: U.S. Department of Health and Human Services, Centers for Disease Control and Prevention, National Institute for Occupational Safety and Health, Fire Fighter Fatality Investigation and Prevention Program Report No. F2002–43 [www.cdc.gov/niosh/fire/reports/face200243.html].

NIOSH [2002b]. Fire fighter dies during live fire training—North Carolina. Morgantown, WV: U.S. Department of Health and Human Services, Centers for Disease Control and Prevention, National Institute for Occupational Safety and Health, Fire Fighter Fatality Investigation and Prevention Program Report No. F2002–19 [www.cdc.gov/niosh/fire/reports/face200219.html].

NIOSH [2002c]. Structural collapse at residential fire claims lives of two volunteer fire chiefs and one career fire fighter—New Jersey. Morgantown, WV: U.S. Department of Health and Human Services, Centers for Disease Control and Prevention, National Institute for Occupational Safety and Health, Fire Fighter Fatality Investigation and Prevention Program Report No. F2002–32 [www.cdc.gov/niosh/fire/reports/face200232.html].

NIOSH [2002d]. First-floor collapse during residential basement fire claims the life of two fire fighters (career and volunteer) and injures a career fire fighter captain—New York. Morgantown, WV: U.S. Department of Health and Human Services, Centers for Disease Control and Prevention, National Institute for Occupational Safety and Health, Fire Fighter Fatality Investigation and Prevention Program Report No. F2002–06 [www.cdc.gov/niosh/fire/reports/face200206.html].

NIOSH [2002e]. Volunteer lieutenant dies following structure collapse at residential house fire—Pennsylvania. Morgantown, WV: U.S. Department of Health and Human Services, Centers for Disease Control and Prevention, National Institute for Occupational Safety and Health, Fire Fighter Fatality Investigation and Prevention Program Report No. F2002–49 [www.cdc.gov/niosh/fire/reports/face200249.html].

NIOSH [2002f]. Structural collapse at an auto parts store fire claims the lives of one career lieutenant and two volunteer fire fighters—Oregon. Morgantown, WV: U.S. Department of Health and Human Services, Centers for Disease Control and Prevention, National Institute for Occupational Safety and Health, Fire Fighter Fatality Investigation and Prevention Program Report No. F2002–50 [www.cdc.gov/niosh/fire/reports/face200250.html].

NIOSH [2003]. Volunteer assistant chief dies in tanker rollover—New Mexico. Morgantown, WV: U.S. Department of Health and Human Services, Centers for Disease Control and Prevention, National Institute for Occupational Safety and Health, Fire Fighter Fatality Investigation and Prevention Program Report No. F2003–23 [www.cdc.gov/niosh/fire/reports/face200323.html].

NIOSH [2007]. Preventing fire fighter fatalities due to heart attacks and other sudden cardiovascular events. Cincinnati, OH: U.S. Department of Health and Human Services, Centers for Disease Control and Prevention, National Institute for Occupational Safety and Health, DHHS (NIOSH) Publication No. 2007–133.

Pelletier KR [2001]. A review and analysis of the clinical- and cost-effectiveness studies of comprehensive health promotion and disease management programs at the worksite: 1998–2000 update. Am J Hlth Promot *16*:107–116.

SRDTF [2008]. Sacramento, CA: Sacramento Regional Driver Training Facility [http://www.srdtf.com/].

Stein AD, Shakour SK, Zuidema RA [2000]. Financial incentives, participation in employer sponsored health promotion, and changes in employee health and productivity: HealthPlus health quotient program. J Occup Environ Med *42*:1148–1155.

Stevens SR, Dempsey WL, Snell CR [2002]. The reduction of occupational absenteeism following two years of fighter wellness program. Med Sci Sports Exercise *34*(5):S1–S194.

USFA [1995]. Fire department communications manual: a basic guide to system concepts and equipment. Emmitsburg, MD: Department of Homeland Security, Federal Emergency Management Agency, U.S. Fire Administration [www.usfa.fema.gov/downloads/pdf/publications/fa-160.pdf].

USFA [1996]. Personnel accountability system technology assessment. Emmitsburg, MD: Department of Homeland Security, Federal Emergency Management Agency, United States Fire Administration, Publication No. EME–96–CO–0587.

USFA [1999a]. Developing effective standard operating procedures: for fire and EMS departments. Emmitsburg, MD: Department of Homeland Security, Federal Emergency Management Agency, U.S. Fire Administration, Publication No. EME–98–CO–0202.

USFA [1999b]. National Fire Academy: managing company tactical operations. Emmitsburg, MD: Department of Homeland Security, Federal Emergency Management Agency, United States Fire Administration.

USFA [2003a]. Improving fire fighter communications: special report. Emmitsburg, MD: Department of Homeland Security, Federal Emergency Management Agency, U.S. Fire Administration, Publication No. EMW–94–C–4423.

USFA [2003b]. Rapid intervention teams: how to avoid needing them: special report. Emmitsburg, MD: Department of Homeland Security, Federal Emergency Management Agency, United States Fire Administration, Publication No. EME–97–CO–0506.

USFA [2004]. Emergency vehicle safety initiative. Emmitsburg, MD: Department of Homeland Security, Federal Emergency Management Agency, U.S. Fire Administration, Publication No. FA–272.

USFA [2006]. Fire data: Fire fighter casualties. Emmitsburg, MD: Department of Homeland Security, Federal Emergency Management Agency, U.S. Fire Administration [www.usfa.dhs.gov/fireservice/fatalities/statistics/casualties.shtm].

USFA [2008]. Emerging Health and Safety Issues in the Voluntary Fire Service. Emmitsburg, MD: Department of Homeland Security, Federal Emergency Management Agency, U.S. Fire Administration, Publication No. FA-317 [http://www.usfa.dhs.gov/downloads/pdf/publications/fa_317.pdf]

Willich SN, Lewis M, Lowel H, Arntz H, Schubert F, Schroder R [1993]. Physical exertion as a trigger of acute myocardial infarction. N Engl J Med 329:1684–1690.

Womack JW, Humbarger CD, Green JS [2005]. Coronary artery disease risk factors in fire fighters: effectiveness of a one year voluntary health and wellness program. Med Sci Sports Exercise 37(5):S1–S385.

Additional NIOSH Fire Fighter Publications

NIOSH [1994]. NIOSH Alert: request for assistance in preventing injuries and deaths of fire fighters. Cincinnati, OH: U.S. Department of Health and Human Services, Centers for Disease Control and Prevention, National Institute for Occupational Safety and Health, DHHS (NIOSH) Publication No. 94–125 [www.cdc.gov/niosh/fire.html].

NIOSH [1997]. Exploding flashlights: are they a serious threat to workers safety? Cincinnati, OH: U.S. Department of Health and Human Services, Centers for Disease Control and Prevention, National Institute for Occupational Safety and Health, DHHS (NIOSH) Publication No. 97–149 [www.cdc.gov/niosh/fact0002.html].

NIOSH [1999]. FDA and NIOSH public health advisory: explosions and fires in aluminum oxygen regulators. Washington, DC: Food and Drug Administration, Center for Devices and Radiological Health [www.fda.gov/cdrh/oxyreg.html].

NIOSH [1999]. NIOSH Alert: preventing injuries and deaths of fire fighters due to structural collapse. Cincinnati, OH: U.S. Department of Health and Human Services, Centers for Disease Control and Prevention National Institute for Occupational Safety and Health, DHHS (NIOSH) Publication No. 99–146 [www.cdc.gov/niosh/99-146.html].

NIOSH [1999]. NIOSH Hazard ID: fire fighting hazards during propane tank fires. Cincinnati, OH: U.S. Department of Health and Human Services, Centers for Disease Control and Prevention, National Institute for Occupational Safety and Health, DHHS (NIOSH) Publication No. Publication No. 99–129 [http://www.cdc.gov/niosh/hid7.html].

NIOSH [2001]. NIOSH Hazard ID: traffic hazards to fire fighters while working along roadways. Cincinnati, OH: U.S. Department of Health and Human Services, Centers for Disease Control and Prevention, National Institute for Occupational Safety and Health, DHHS (NIOSH) Publication No. 2001–143 [www.cdc.gov/niosh/hid12.html].

NIOSH [2001]. NIOSH Health Hazard Evaluation Report: Evaluation of Woodland Plastics Recycling Corporation Incident—Madison, WI. Cincinnati, OH: U.S. Department of Health and Human Services, Centers for Disease Control and Prevention, National Institute for Occupational Safety and Health, DHHS (NIOSH) Publication No. HETA 2001–0043–2844 [www.cdc.gov/niosh/hhe/reports/pdfs/2001-0043-2844.pdf].

NIOSH [2002]. Fire fighter deaths from tanker truck rollovers. Cincinnati, OH: U.S. Department of Health and Human Services, Centers for Disease Control and Prevention, National Institute for Occupational Safety and Health, DHHS (NIOSH) Publication No. 2002–111 [www.cdc.gov/niosh/hid14.html].

NIOSH [2002]. NIOSH Hazard ID: fire fighters exposed to electrical hazards during wildland fire operations. Cincinnati, OH: U.S. Department of Health and Human Services, Centers for Disease Control and Prevention, National Institute for Occupational Safety and Health, DHHS (NIOSH) Publication No.2002–112 [http://www.cdc.gov/niosh/hid15.html].

NIOSH [2003]. Fire fighter fatality/injury reports and other related publications CD-ROM. Cincinnati, OH: U.S. Department of Health and Human Services, Public Health Service, Centers for Disease Control and Prevention, National Institute for Occupational Safety and Health, DHHS (NIOSH) Publication No. 2003-136.

NIOSH [2003]. Your safety 1st—railroad crossing safety for emergency responders. Cincinnati, OH: U.S. Department of Health and Human Services, Centers for Disease Control and Prevention, National Institute for Occupational Safety and Health, DHHS (NIOSH) Publication No. 2003–121 [www.cdc.gov/niosh/docs/2003-121/].

NIOSH [2004]. Protecting emergency responders. Volume 3: Safety management in disaster and terrorism response. Cincinnati, OH: U.S. Department of Health and Human Services,

Centers for Disease Control and Prevention, National Institute for Occupational Safety and Health, DHHS (NIOSH) Publication No. 2004–144 [www.cdc.gov/niosh/docs/2004-144/].

NIOSH [2004]. Workplace Solutions: divers beware: training dives present serious hazards to fire fighters. Cincinnati, OH: U.S. Department of Health and Human Services, Centers for Disease Control and Prevention, National Institute for Occupational Safety and Health, DHHS (NIOSH) Publication No. 2004–152 [www.cdc.gov/niosh/docs/wp-solutions/2004-152/].

NIOSH [2005]. NIOSH Alert: preventing injuries and deaths of fire fighters due to truss system failures. Cincinnati, OH: U.S. Department of Health and Human Services, Centers for Disease Control and Prevention, National Institute for Occupational Safety and Health, DHHS (NIOSH) Publication No. 2005–132 [www.cdc.gov/niosh/docs/2005-132/].

NIOSH [2005]. Workplace Solutions—preventing deaths and injuries to fire fighters during live-fire training in acquired structures. Cincinnati, OH: U.S. Department of Health and Human Services, Centers for Disease Control and Prevention, National Institute for Occupational Safety and Health, DHHS (NIOSH) Publication No. 2005–102 [www.cdc.gov/niosh/docs/wp-solutions/2005-102/].

NIOSH [2006]. FDA and NIOSH Public Health Notification: oxygen regulator fires resulting from incorrect use of CGA 870 seals [www.fda.gov/cdrh/safety/042406-o2fires.html].

NIOSH [2007]. Preventing fire fighter fatalities due to heart attacks and other sudden cardiovascular events. Cincinnati, OH: U.S. Department of Health and Human Services, Centers for Disease Control and Prevention, National Institute for Occupational Safety and Health, DHHS (NIOSH) Publication No. 2007–133 [www.cdc.gov/niosh/docs/2007-133/].

NIOSH [ongoing]. NIOSH Health Hazard Evaluations (HHEs). Cincinnati, OH: U.S. Department of Health and Human Services, Centers for Disease Control and Prevention, National Institute for Occupational Safety and Health. (To date, 37 fire fighter HHEs have been completed and are available at www.cdc.gov/niosh/hhe/; search word is fire fighter.)

Questions?

If you have questions regarding the NIOSH FFFIPP, please contact the NIOSH Division of Safety Research at

> National Institute for Occupational Safety and Health
> Division of Safety Research
> Surveillance and Field Investigations Branch
> 1095 Willowdale Road, M/S H–1808
> Morgantown, WV 26505–2888
>
> Phone: 304–285–5916
> FAX: 304–285–5774
> www.cdc.gov/niosh/im-dsr.html

www.ingramcontent.com/pod-product-compliance
Lightning Source LLC
Chambersburg PA
CBHW081905170526
45167CB00007B/3157